KnotMonsters

Biology edition
20 Amigurumi Crochet Patterns
by Michael Cao

KnotMonsters

Copyright © 2021 Michael Cao (@KnotMonster), Photography © 2021 Sushi Aquino

All rights reserved. All rights reserved. No part of this publication may be reproduced, stored in a retrieval system or transmitted in any form or by any means without the prior written permission of the publisher and copyright owner.

The book is sold subject to the condition that all designs are copyright and are not for commercial reproduction without the permission of the designer and copyright owner.

**To those who build satisfaction with their hands
and happiness with their heart**

Nothing is hard, only new	1
Hook, yarn and sinker	3
How to crochet	7
Spirilla	21
Bacilla	23
Cocci ver 1	25
Cocci ver 2	27
Vibrio ver 1	29
Vibrio ver 2	31
Helicobacter	33
Corynebacteria ver 1	35
Corynebacteria ver 2	37
Escherichia ver 1	39
Escherichia ver 2	41
Bifidobacterium	43
Microscope	45
Animal Eukaryotic Cell	51
Plant Eukaryotic Cell	58
Virus - bacteriophage	65
DNA	69
Virus – COVID 19	72
XY Chromosomes	75

What started as something to pass the time during quarantine has turned into so much more. These toys bring me joy when I make them and even more when I give them away. I hope that my creations will bring you and your lucky loved ones hours of happiness as well. I want to give a special thanks to *YOU* for making my creations and bringing them to life. Your continued support encourages my creative side.

THANK YOU

NOTHING IS HARD, ONLY NEW

If you have never crocheted before, **NO PROBLEM**. Everyone has to start somewhere. When I first started, it took me a week just to figure out how to make a ball. It took even longer to get my fingers and hands used to holding a crochet hook. Crocheting can be frustrating at first but like with all things new, your body will develop muscle memory and over time, you will find that it will get easier. Always remember that nothing is hard, only new. There is no right way or wrong way to crochet. Whatever works for you, your body, and your ability is the right way. Some people will hold a crochet hook like a pen, I prefer to hold it like a knife, with my index finger along the length of the hook. This is purely personal preference. Crochet is an art form and like all art forms, is open to personal interpretation. Crochet patterns function as guidelines only, and I encourage you to bring your own artistic flair to each pattern. Do not be afraid to use different hooks or types of yarn. Experiment by changing the colors. Add accessories, modify the body shape, and bring your own individual spice to each pattern. Sometimes it will work; most of the time, it will not, and that is perfectly fine because in the end, you will end up with something that is yours. Something that you created. Your own little knotmonster.

1

HOOK, YARN, AND SINKER

Crocheting requires two fundamental things: a hook and yarn. Hook sizes are measured in millimeters and are also given a corresponding letter. For example, a 3.75mm hook is called an "F" hook and a 4mm hook is called a "G" hook. For amigurumi, I like to use a 3.75mm or a 4mm hook with worsted weight yarn. The smaller the hook, the tighter the stitches will be; the bigger the hook, the looser the stitches will be. The yarn is complementary to the hook. A thinner yarn will require a smaller hook, and a thicker yarn will require a larger one. Let's talk about what else you will need.

To start, you will need:

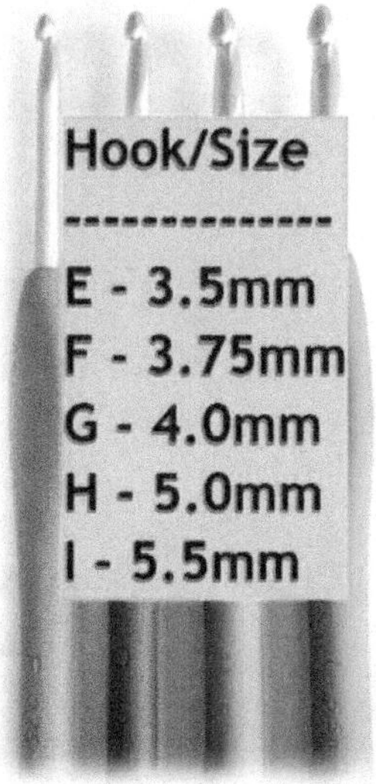

Hook [3.75mm (F) or 4mm (G)]

Yarn (worsted weight)

Scissors

Stitch marker (eg. scrap yarn, paper clip)

Safety eyes (8-14 mm)

Yarn needle #16 with blunt tip

Stuffing material

Pen/paper

Additional items:

Ruler

Hot glue or fabric glue

Felt (multiple colors)

Embroidery thread

Fabric scissors are specifically designed to cut fabric. They are incredibly sharp and will cut the yarn without fraying of the ends. This makes it much easier to thread the yarn through the needle when you are tying off.

Safety eyes are called "safety" because they have a small hole in the bottom of the eye. If a small child accidentally aspirates the eye, this hole will allow for air to pass through it. This is the concept, but it is not fool proof. Always use caution when you are giving a toy to a child under the age of three or to a child who tends to put things in their mouth. Alternatively, for these children, I would highly recommend making eyes out of yarn or felt instead. It is also important to thoroughly tie in your loose ends so your toys will not unravel and fall apart.

Stitch markers can be purchased from any craft store or online. Stitch markers help you keep track of where your first stitch in a row is. By marking your first stitch with a stitch marker, you will know when you have reached the end of your row. Alternatively, you can use a paper clip or a piece of scrap yarn (my personal favorite).

You want a needle that is wide enough to allow for worsted weight yarn to pass through the eye, while small enough to comfortably weave in your ends when you finish. A trick to easily thread your needle is to flatten the end of the yarn.

Accessories and pieces can be sewn or glued to your knotmonsters. Whenever possible, sew the pieces on, as they will hold much better than glue. However, glue works great to attach felt eyes as long as they are thoroughly attached, including the edges. When sewing, I recommend using the mattress stitch technique in order to hide the stitch as much as possible. For a mattress stitch, draw the yarn up one stitch and back down through the next stitch on the same piece. Then attach it to the other piece and repeat

2

HOW TO CROCHET

When you start a new pattern, the technique of crocheting will be noted as either completed in the straight or the round. Straight crocheting is completed in rows by going to the end of a row and turning around, creating a square-like pattern. Round crocheting is completed in rounds and, starting at the center, is done in spirals creating a circular pattern. Typically, a chain is used to start a straight crochet while a magic ring is used to start a round crochet. Exceptions to this are when an oval shape is desired. In this situation, you may form a chain and then, instead of turning at the end of the row, continuing onwards to the opposite side of the chain, thus creating an oval.

Patterns are written in different ways around the world. For example, in a UK pattern, a double crochet can mean a single crochet in a US pattern. For the sake of simplicity, all my patterns are based in US format and I have done my best to make them as simple and easy to read as possible. Each row/round is delineated by "R1, R2, R3, etc." When crocheting in the straight, at the end of each row, chain 1, turn the piece and start on the next row in the second chain from your hook. This step is necessary to create height in the piece. When crocheting in the round, one round means that you have crocheted completely around and have returned to the first stitch that you placed from the previous round. No chain/turning is necessary when crocheting in the round using any of my patterns because we will be working in a spiral motion. At the end of each row/round you will see a (#). This number represents how many stitches you should end up with when you finish the row/round.

Let's do an example: "R3: (SC, inc) x 6 (18)"

This means that we are on round number three, and we will do a single crochet and then an increase; and then repeat that five more times. At the end of our round, we will end up with eighteen total stitches.

Difficulty is rated by easy, medium, or advanced, based on the number of unique stitch techniques in each pattern.

While standard crochet will work just fine for amigurumi patterns, I use cross stitching for all my amigurumi. Cross stitching tightens up the stitches and helps prevent the stuffing from showing and the stitches from stretching. It also makes a neat grid-like texture. Therefore, <u>the following tutorials are for cross stitching and not standard crochet</u>. If you are familiar with crocheting and notice that these instructions are backwards by yarning over instead of under, it is because of cross stitching. The patterns will look perfectly adorable if you decide to not use cross stitching.

Slipknot

A slip knot allows you to tighten the knot with the loose yarn end. To create a slipknot, pull a loop made of the loose yarn end through the top of the first loop and tighten the knot. Pull on the loose yarn end to tighten the loop.

Chain (ch)

Start with your hook around a slipknot. Yarn over (yarn is over the hook) and pull through. You have made one chain or "ch 1." To create multiple chains, continue yarning over until desired chain # is met.

Slip Stitch (sl)

Enter next stitch with hook. Yarn under (yarn is under the hook – note this is for cross stitch) and pull through both loops. When you get to future rows, for sl, SC, DC, and TC, be sure to insert your hook under both loops when you start.

Single Crochet (SC)

Enter next stitch with hook. Yarn under (yarn is under the hook – note this is for cross stitch) and pull through first loop. Yarn over, pull through both loops.

Half Double Crochet (HDC)

Yarn over, enter next stitch with hook. Yarn under (yarn is under the hook – note this is for cross stitch) and pull through all three loops.

Double Crochet (DC)

Yarn over, enter next stitch with hook. Yarn under (yarn is under the hook – note this is for cross stitch) and pull through first two loops. Yarn over and pull through last two loops.

Triple Crochet (TC)

Yarn over twice, enter next stitch with hook. Yarn under (yarn is under the hook – note this is for cross stitch) and pull through first two loops. Yarn over and pull through next two loops. Yarn over and pull through last two loops.

Increase (inc)

Single crochet twice in the same stitch.

Invisible Decrease (dec)

While not completely invisible and the name may suggest, the goal of an invisible decrease is to reduce the number of stitches in a row while creating minimal seams. This is done by inserting the hook into the next stitch. Then insert the hook into the following stitch. These stitches will be the front loops facing you on subsequent rows. Yarn under (note this is for cross stitch) and pull through first two loops. Then yarn over and pull through both loops.

Magic Ring

Position fingers and yarn as shown in (1), grab yarn from underneath loose end (2), twist (3), yarn over (4), pull through (5). You will not have a slipknot on a loop, so if you pull on the loose end the loop will close. Single crochet six times into the ring (9) and tighten up the loop (10). This will create a magic ring 6 (MR 6) which is the most common start to most amigurumi.

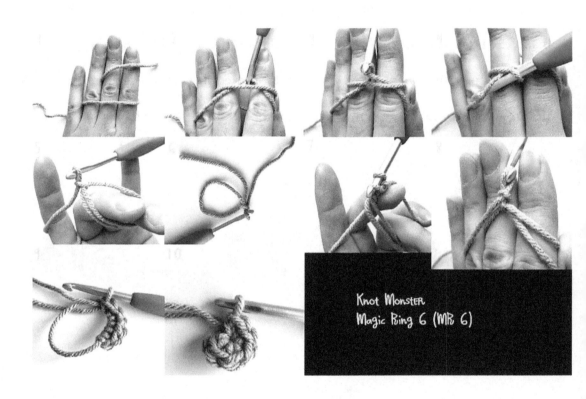

Front loops only (FLO) & Back loops only (BLO)

When a pattern calls for FLO or BLO, this means that you will be doing the next row in the back or the front loops only. If the next row does not indicate FLO/BLO, then switch back to normal crochet under both loops. Back and front is relative to you, so if the loop is facing you then that is the front loop and vice versa.

Switching colors

To change colors, when you reach the next row, draw up the new color through both loops and then single crochet into the next stitch. Tie both loose ends together in the back of the project.

Puff Stitch

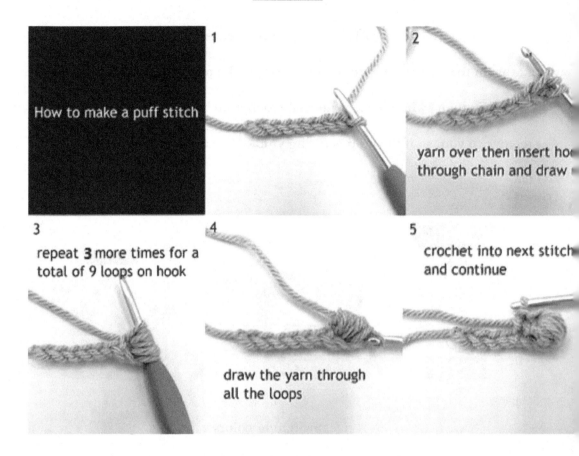

Oval Shape

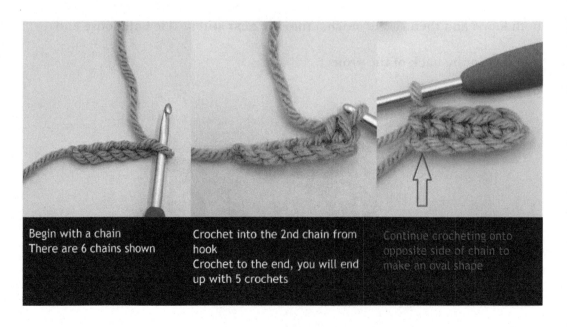

Attaching two feet together

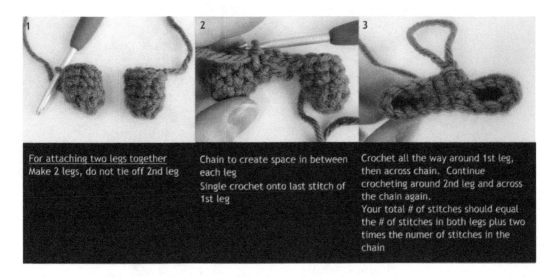

For attaching two legs together
Make 2 legs, do not tie off 2nd leg

Chain to create space in between each leg
Single crochet onto last stitch of 1st leg

Crochet all the way around 1st leg, then across chain. Continue crocheting around 2nd leg and across the chain again.
Your total # of stitches should equal the # of stitches in both legs plus two times the numer of stitches in the chain

Tying off

When you reach the end of your project. To tie off, slip stitch into the next stitch, then chain one. Pull the yarn all the way through, leaving a long end for sewing and cut the yarn where the red arrow is. Thread the yarn through a #16 yarn needle and weave it through several nearby stitches. Do this in three separate directions and then pass the yarn through the project and cut the yarn to hide the loose end.

Bacteria Babies

Difficulty: Medium

Hook: 4.0mm (G) or 3.75 mm (F) hook

Approximate size: 10cm x 10cm (varies)

Eyes: 6 mm

<u>**All rows completed in the round unless otherwise indicated**</u>

Spirilla

Start with color pink

R1: MR 6 (6)

R2: inc x 6 (12)

R3: (SC, inc) x 6 (18)

R4-8: SC 18 (18)

As you continue the pattern, stuff as you go. Place eyes, and use embroidery thread for cheeks.

R9-33: (SC, dec) x 3, (SC 2, inc) x 3 (18)

Stuff

R34: (SC, dec) x 6 (12)

R35: dec x 6 (6)

Tie off

Bacilli

Start with color green

R1: MR 6 (6)

R2: inc x 6 (12)

R3: (SC, inc) x 6 (18)

R4: (SC 2, inc) x 6 (24)

R5-16: SC 24 (24)

Place eyes, and use embroidery thread for cheeks. Stuff.

R17: (SC 2, dec) x 6 (18)

R18: dec x 9 (9)

R19: inc x 9 (18)

R20: (SC 2, inc) x 6 (24)

R21-32: SC 24 (24)

Stuff

R33: (SC 2, dec) x 6 (18)

R34: (SC, dec) x 6 (12)

R35: dec x 6 (6)

Tie off

Cocci ver 1

Start with color blue

R1: MR 6 (6)

R2: inc x 6 (12)

R3: (SC, inc) x 6 (18)

R4: (SC 2, inc) x 6 (24)

R5-7: SC 24 (24)

R8: (SC 2, dec) x 6 (18)

Place eyes, and use embroidery thread for cheeks. Stuff.

R9: (SC, dec) x 6 (12)

R10: dec x 6 (6)

R11: inc x 6 (12)

R12: (SC, inc) x 6 (18)

R13: (SC 2, inc) x 6 (24)

R14-16: SC 24 (24)

R17: (SC 2, dec) x 6 (18)

Stuff

R18: (SC, dec) x 6 (12)

R19: dec x 6 (6)

R20: inc x 6 (12)

R21: (SC, inc) x 6 (18)

R22: (SC 2, inc) x 6 (24)

R23-25: SC 24 (24)

R26: (SC 2, dec) x 6 (18)

Stuff

R27: (SC, dec) x 6 (12)

R28: dec x 6 (6)

R29: inc x 6 (12)

R30: (SC, inc) x 6 (18)

R31: (SC 2, inc) x 6 (24)

R32-34: SC 24 (24)

R35: (SC 2, dec) x 6 (18)

Stuff

R36: (SC, dec) x 6 (12)

R37: dec x 6 (6)

Tie off

Cocci ver 2

Start with color light blue

R1: MR 6 (6)

R2: inc x 6 (12)

R3: (SC, inc) x 6 (18)

R4: (SC 2, inc) x 6 (24)

R5: (SC 3, inc) x 6 (30)

R6: (SC 4, inc) x 6 (36)

R7-11: SC 36 (36)

Place eyes, and use embroidery thread for cheeks.

R12: (SC 4, dec) x 6 (30)

R13: (SC 3, dec) x 6 (24)

Stuff

R14: (SC 2, dec) x 6 (18)

R15: dec x 6 (9)

R16: inc x 6 (18)

R17: (SC 2, inc) x 6 (24)

R18: (SC 3, inc) x 6 (30)

R19: (SC 4, inc) x 6 (36)

R20-24: SC 36 (36)

R25: (SC 4, dec) x 6 (30)

R26: (SC 3, dec) x 6 (24)

R27: (SC 2, dec) x 6 (18)

Stuff

R28: (SC, dec) x 6 (12)

R29: dec x 6 (6)

Tie off

Vibrio ver 1

Start with color purple

R1: MR 6 (6)

R2: inc x 6 (12)

R3: (SC, inc) x 6 (18)

R4: (SC 2, inc) x 6 (24)

R5: (SC 3, inc) x 6 (30)

R6: SC 30 (30)

R7: inc, SC 14, dec, SC 13 (30)

R8: SC 30 (30)

R9: inc, SC 14, dec, SC 13 (30)

R10: SC 30 (30)

R11: inc, SC 14, dec, SC 13 (30)

Place eyes, and use embroidery thread for cheeks.

R12-14: SC 30 (30)

R15: dec, SC 14, inc, SC 13 (30)

R16: SC 30 (30)

R17: dec, SC 14, inc, SC 13 (30)

R18: SC 30 (30)

R19: dec, SC 14, inc, SC 13 (30)

R20: SC 30 (30)

Stuff

R21: (SC 3, dec) x 6 (24)

R22: (SC 2, dec) x 6 (18)

Stuff

R23: (SC, dec) x 6 (12)

R24: dec x 6 (6)

R25: (SC, dec) x 2 (4)

R26-33: SC 4 (4)

Tie off

As you continue the pattern, stuff as you go. Place eyes, and use embroidery thread for cheeks.

R13: (SC 2, dec) x 6 (18)

R14-16: SC 18 (18)

R17: (SC, dec) x 6 (12)

R18-20: SC 12 (12)

R21: dec x 6 (6)

R22: (SC, dec) x 2 (4)

R23-29 SC 4 (4)

Tie off

Vibrio ver 2

Start with color light green

R1: MR 6 (6)

R2: inc x 6 (12)

R3: (SC, inc) x 6 (18)

R4: SC 18 (18)

R5: (SC 2, inc) x 6 (24)

R6: (SC 3, inc) x 6 (30)

R7-9: SC 30 (30)

R10: (SC 3, dec) x 6 (24)

R11-12: SC 24 (24)

R15-19: inc, SC 8, dec, SC 7 (18)

Stuff

R20: (SC, dec) x 6 (12)

R21: dec x 6 (6)

Tie off

Part 2 (make 3)

Ch 19, turn, starting in 2nd chain from hook: (inc, SC 2) x 6

Attach to bottom of part 1

Helicobacter

Part 1

Start with color orange

R1: MR 6 (6)

R2: inc x 6 (12)

R3: (SC, inc) x 6 (18)

R4: SC 18 (18)

R5-9: inc, SC 8, dec, SC 7 (18)

Place eyes, and use embroidery thread for cheeks. Stuff.

R10-14: dec, SC 7, inc, SC 8 (18)

Stuff

Corynebacteria ver 1

Start with color yellow

R1: MR 6 (6)

R2: inc x 6 (12)

R3: (SC, inc) x 6 (18)

R4: (SC 2, inc) x 6 (24)

R5: (SC 3, inc) x 6 (30)

R6-9: SC 30 (30)

R10: (SC 3, dec) x 6 (24)

Place eyes, and use embroidery thread for cheeks.

R11: (SC 2, dec) x 6 (18)

R12: FLO SC 18 (18)

R13-22: SC 18 (18) -stuff as you go

R23: (SC, dec) x 6 (12)

R24: dec x 6 (6)

Tie off

Corynebacteria ver 2

Start with color yellow

R1: MR 6 (6)

R2: inc x 6 (12)

R3: (SC, inc) x 6 (18)

R4: (SC 2, inc) x 6 (24)

R5: (SC 3, inc) x 6 (30)

R6-9: SC 30 (30)

R10: (SC 3, dec) x 6 (24)

R11: (SC 2, dec) x 6 (18)

R12: FLO SC 18 (18)

R13-22: SC 18 (18) - Stuff as you go.

Place eyes, and use embroidery thread for cheeks.

R23: FLO (SC 2, inc) x 6 (24)

R24: (SC 3, inc) x 6 (30)

R25-28: SC 30 (30)

R29: (SC 3, dec) x 6 (24)

R30: (SC 2, dec) x 6 (18)

Stuff

R31: (SC, dec) x 6 (12)

R32: dec x 6 (6)

Tie off

Escherichia ver 1

Start with color gray

Ch 11, turn and begin in 2nd chain from hook

R1: SC 10, continue onto other side of chain SC 10 (20)

R2: inc, SC 8, inc x 2, SC 8, inc (24)

R3: SC, inc, SC 8, (SC, inc) x 2, SC 9, inc (28)

R4: (SC, inc) x 2, SC 8, (SC, inc) x 3, SC 9, inc (34)

R5: (SC 2, inc) x 2, SC 8, (SC 2, inc) x 3, SC 10, inc (40)

R6-11: SC 40 (40)

R12: (SC 2, dec) x 2, SC 8, (SC 2, dec) x 3, SC 9, dec (34)

R13: (SC, dec) x 2, SC 8, (SC, dec) x 3, SC 9, dec (28)

R14: SC, dec, SC 8, (SC, dec) x 2, SC 9, dec (28)

Place eyes, and use embroidery thread for cheeks.

R15: dec, SC 8, dec x 2, SC 8, dec (24)

Stuff, sew closed

Attach pieces of tan colored yarn by tying a knot around each stitch. Continue all the way around the body. Trim yarn to desired length.

I've got your bac

Escherichia ver 2

Start with color pink

Ch 11, turn and begin in 2nd chain from hook

R1: SC 10, continue onto other side of chain SC 10 (20)

R2: inc, SC 8, inc x 2, SC 8, inc (24)

R3: SC, inc, SC 8, (SC, inc) x 2, SC 9, inc (28)

R4: (SC, inc) x 2, SC 8, (SC, inc) x 3, SC 9, inc (34)

R5: (SC 2, inc) x 2, SC 8, (SC 2, inc) x 3, SC 10, inc (40)

R6-11: SC 40 (40)

R12: (SC 2, dec) x 2, SC 8, (SC 2, dec) x 3, SC 9, dec (34)

R13: (SC, dec) x 2, SC 8, (SC, dec) x 3, SC 9, dec (28)

R14: SC, dec, SC 8, (SC, dec) x 2, SC 9, dec (28)

Place eyes, and use embroidery thread for cheeks.

R15: dec, SC 8, dec x 2, SC 8, dec (24)

Stuff, sew closed

Part 2

Ch 55. Turn, starting in 2nd chain from hook: inc x 54 (108). Attach all the way around the body. You may need to modify the length based on your tension.

R26: dec x 6 (6)

Tie off

We are now going to continue by crocheting the skipped stitches from R20

Bifidobacterium

Start with color purple

R1: MR 6 (6)

R2: inc x 6 (12)

R3-16: SC 12 (12) - stuff as you go

R17: inc x 12 (24)

R18-19: SC 24 (24)

Place eyes, and use embroidery thread for cheeks.

R20: Skip 12, SC 12 (12)

We will be working on one half of the pattern first.

R21-25: SC 12 (12) -stuff as you go

Microscope

Difficulty: Medium

Hook: 4.0mm (G) or 3.75 mm (F) hook

Approximate size: 18cm x 10cm

All rows completed in the round unless otherwise indicated

BASE

Start with color grey

"(SC 3) = (3 single crochets in 1 stitch)"

R1: MR 8 (8)

R2: [(SC 3), SC] x 4 (16)

R3: [(SC 3), SC 3] x 4 (24)

R4: [(SC 3), SC 5] x 4 (32)

R5: [(SC 3), SC 7] x 4 (40)

R6: [(SC 3), SC 9] x 4 (48)

R7: BLO SC 48 (48)

R8-10: SC 48 (48)

Change color to black

R11: BLO (dec, SC 10) x 4 (44)

R12: (dec, dec, SC 7) x 4 (36)

R13: (dec, SC 5, dec) x 4 (28)

R14: (dec, SC 3, dec) x 4 (20)

Stuff lightly

R15: (dec, SC, dec) x 4 (12)

Change color to white

R16: dec x 6 (6)

Tie off

BASE CLIPS (make 2)

Start with color grey

Ch 5, turn, starting in 2nd chain from hook: sl 4, tie off and attach to base

STAND

Start with color grey

R1: MR 6 (6)

R2: inc x 6 (12)

R3: (SC, inc) x 6 (18)

R4: BLO SC 18 (18)

R5-24: SC 18 (18) - stuff as you go

R25: (SC, dec) x 6 (12)

R26: dec x 6 (6)

Tie off, attach to base

EYEPIECE TUBE

Start with color black

R1: MR 6 (6)

R2: inc x 6 (12)

R3: BLO SC 12 (12)

R4-8: SC 12 (12)

Change color to grey

R9: FLO inc x 12 (24)

R10-27: SC 24 (24) – stuff as you go

Change color to black

R28: FLO (SC 3, inc) x 6 (30)

R29: BLO (SC 3, dec) x 6 (24)

Stuff

R30: (SC 2, dec) x 6 (18)

R31: (SC, dec) x (12)

R32: dec x 6 (6)

Tie off, attach to stand

OBJECTIVE LENSES

Part 1

Start with color black

R1: MR 6 (6)

R2: inc x 6 (12)

R3: BLO SC 12 (12)

Change color to red

R4: SC 12 (12)

Change color to black

R5: SC 12 (12)

Stuff, tie off, and attach to tube

Part 2

Start with color black

R1: MR 6 (6)

R2: inc x 6 (12)

R3: BLO SC 12 (12)

Change color to yellow

R4: SC 12 (12)

Change color to black

R5-6: SC 12 (12)

Stuff, tie off, and attach to tube

Part 3

Start with color black

R1: MR 6 (6)

R2: inc x 6 (12)

R3: BLO SC 12 (12)

Change color to blue

R4: SC 12 (12)

Change color to black

R5-7: SC 12 (12)

Stuff, tie off, and attach to tube

ADJUSTMENT KNOBS (make 4)

Start with color black

R1: MR 6 (6)

R2: inc x 6 (12)

R3: BLO SC 12 (12)

Stuff, tie off, attach

Animal Eukaryotic Cell Miniature Pillow

Difficulty: Medium

Hook: 4.0mm (G) or 3.75 mm (F) hook

Approximate size: 20cm x 10cm

<u>All rows completed in the round unless otherwise indicated</u>

BODY

Part 1

Start with color light brown

*HDC inc = 2 HDC's in 1 stitch

R1: 10 HDC's in a MR (10)

R2: HDC inc x 10 (20)

R3: (HDC, HDC inc) x 10 (30)

R4: (HDC 2, HDC inc) x 10 (40)

R5: (HDC 3, HDC inc) x 10 (50)

R6: (HDC 4, HDC inc) x 10 (60)

R7: (HDC 5, HDC inc) x 10 (70)

R8: (HDC 6, HDC inc) x 10 (80)

R9: (HDC 7, HDC inc) x 10 (90)

R10: (HDC 8, HDC inc) x 10 (100)

Tie off

Part 2

Start with color brown

*HDC inc = 2 HDC's in 1 stitch

R1: 10 HDC's in a MR (10)

R2: HDC inc x 10 (20)

R3: (HDC, HDC inc) x 10 (30)

R4: (HDC 2, HDC inc) x 10 (40)

R5: (HDC 3, HDC inc) x 10 (50)

R6: (HDC 4, HDC inc) x 10 (60)

R7: (HDC 5, HDC inc) x 10 (70)

R8: (HDC 6, HDC inc) x 10 (80)

R9: (HDC 7, HDC inc) x 10 (90)

R10: (HDC 8, HDC inc) x 10 (100)

R11-16: HDC 100 (100)

Attach part 1 to part 2 by single crocheting.

Nucleus part 1

Start with color purple

R1: MR 6 (6)

R2: inc x 6 (12)

Tie off and attach to nucleus part 2

Nucleus part 2

Start with color light purple

R1: MR 6 (6)

R2: inc x 6 (12)

R3: (SC, inc) x 6 (18)

R4: (SC 2, inc) x 6 (24)

R5: (SC 3, inc) x 6 (30)

Change color to blue

R6: (SC 4, inc) x 6 (36)

Tie off, leave end long for attaching to body part 1

ORGANELLES

Organelles are attached to part 1 before part 1 is attached to part 2.

Endoplasmic Reticulum

Start with color blue

Ch 71, turn and begin in 2nd chain from hook, SC 70 (70)

Tie off, leave end very long. Tie them onto the body in a random pattern around edge of nucleus.

Ribosomes (make 10)

Start with color pink

Ch 2, turn and begin in 2nd chain from hook, SC 1 (1)

Tie off, attach 7 to endoplasmic reticulum and 3 to body.

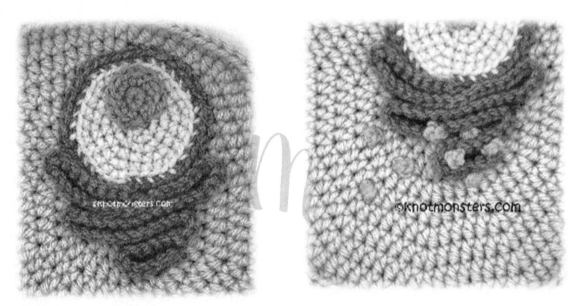

Cell-ebrate good times!

Golgi Apparatus

Part 1

Start with color orange

R1: MR 4 (4)

R2-8: SC 4 (4)

Tie off

Part 2

Start with color orange

R1: MR 4 (4)

R2-7: SC 4 (4)

Tie off

Part 3

Start with color orange

R1: MR 4 (4)

R2-5: SC 4 (4)

Tie off

Attach parts 1, 2 and 3 to body

Perioxisome (make 2)

Start with color light green

R1: MR 6 (6)

Tie off, attach to body

Lysosome (make 2)

Start with color bright pink

R1: MR 6 (6)

R2: inc x 6 (12)

Tie off, attach to body

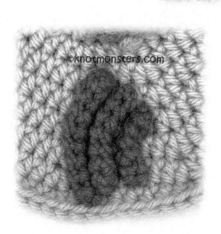

Vacuole (make 2)

Start with color light blue

In a magic ring: TC 3, DC 6, TC 3, tighten magic ring slightly, finish off by slip stitching into magic ring. Tighten magic ring and attach to body

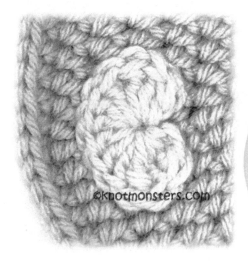

Mitochondria (make 2)

Part 1

Start with color dark yellow

Ch 7, turn and begin in 2nd chain from hook

R1: SC 6, continue onto other side of chain SC 6 (12)

R2: inc, SC 4, inc x 2, SC 4, inc (16)

R3: inc x 2, SC 5, inc x 3, SC 5, inc (22)

Tie off

Part 2

Start with color yellow

Ch 18, tie and attach to part 1. Attach both parts to body.

Attachment

Single crochet body part 2 onto body part 1. Stuff before closing.

Plant Eukaryotic Cell Miniature Pillow

Difficulty: Medium

Hook: 4.0mm (G) or 3.75 mm (F) hook

Approximate size: 20cm x 10cm

<u>All rows completed in the round unless otherwise indicated</u>

BODY

Part 1

Start with color light green

*(HDC 3) = 3 HDC's in 1 stitch

R1: 8 HDC's in a MR (8)

R2: [(HDC 3), HDC] x 4 (16)

R3: [(HDC 3), HDC 3] x 4 (24)

R4: [(HDC 3), HDC 5] x 4 (32)

R5: [(HDC 3), HDC 7] x 4 (40)

R6: [(HDC 3), HDC 9] x 4 (48)

R7: [(HDC 3), HDC 11] x 4 (56)

R8: [(HDC 3), HDC 13] x 4 (64)

R9: [(HDC 3), HDC 15] x 4 (72)

R10: [(HDC 3), HDC 17] x 4 (80)

Tie off

Part 2

Start with color green

*(HDC 3) = 3 HDC's in 1 stitch

R1: 8 HDC's in a MR (8)

R2: [(HDC 3), HDC] x 4 (16)

R3: [(HDC 3), HDC 3] x 4 (24)

R4: [(HDC 3), HDC 5] x 4 (32)

R5: [(HDC 3), HDC 7] x 4 (40)

R6: [(HDC 3), HDC 9] x 4 (48)

R7: [(HDC 3), HDC 11] x 4 (56)

R8: [(HDC 3), HDC 13] x 4 (64)

R9: [(HDC 3), HDC 15] x 4 (72)

R10: [(HDC 3), HDC 17] x 4 (80)

R11: BLO HDC 80 (80)

R12-15: HDC 80 (80)

Nucleus part 1

Start with color purple

R1: MR 6 (6)

R2: inc x 6 (12)

Tie off and attach to nucleus part 2

Nucleus part 2

Start with color light purple

R1: MR 6 (6)

R2: inc x 6 (12)

R3: (SC, inc) x 6 (18)

R4: (SC 2, inc) x 6 (24)

R5: (SC 3, inc) x 6 (30)

Change color to blue

R6: BLO (SC 4, inc) x 6 (36)

Tie off, leave end long for attaching to body part 1

ORGANELLES

Organelles are attached to part 1 before part 1 is attached to part 2.

Endoplasmic Reticulum

Start with color blue

Ch 71, turn and begin in 2nd chain from hook, SC 70 (70)

Tie off, leave end very long. Tie them onto the body in a random pattern around edge of nucleus.

Ribosomes (make 10)

Start with color pink

Ch 2, turn and begin in 2nd chain from hook, SC 1 (1)

Tie off, attach 7 to endoplasmic reticulum and 3 to body.

Let's Cell-ebrate with a cell-fie!

Golgi Apparatus

Part 1

Start with color orange

R1: MR 4 (4)

R2-8: SC 4 (4)

Tie off

Part 2

Start with color orange

R1: MR 4 (4)

R2-7: SC 4 (4)

Tie off

Part 3

Start with color orange

R1: MR 4 (4)

R2-5: SC 4 (4)

Tie off

Attach parts 1, 2 and 3 to body

Vacuole

Start with color light blue

TC inc = 2 TC's in 1 st

DC inc = 2 DC's in 1 st

R1: in a magic ring: TC 3, DC 6, TC 3 (12)

R2: ch 1, turn, starting in 2nd chain from hook: TC inc x 3, DC inc x 6, TC inc x 3, slip stitch into magic ring, tighten magic ring (23)

Tie off, attach to body

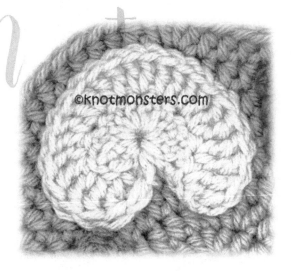

Mitochondria

Part 1

Start with color dark yellow

Ch 7, turn and begin in 2ⁿᵈ chain from hook

R1: SC 6, continue onto other side of chain SC 6 (12)

R2: inc, SC 4, inc x 2, SC 4, inc (16)

R3: inc x 2, SC 5, inc x 3, SC 5, inc (22)

Tie off

Part 2

Start with color yellow

Ch 18, tie and attach to part 1. Attach both parts to body.

Chloroplast

Start with color green

Ch 7, turn and begin in 2ⁿᵈ chain from hook

R1: SC 6, continue onto other side of chain SC 6 (12)

R2: inc, SC, inc x 2, SC, inc x 2, SC, dec, SC, inc (17)

Tie off, attach dark green yarn for accents

Attachment

Single crochet body part 2 onto body part 1. Stuff before closing.

Virus - Bacteriophage

Difficulty: Advanced

Hook: 4.0mm (G) or 3.75 mm (F) hook

Approximate size: 15cm x 15cm

Eye: 6 mm

All rows completed in the round unless otherwise indicated

HEAD/BODY

Start with color light purple

R1: MR 4 (4)

R2: inc x 4 (8)

R3: (SC, inc) x 4 (12)

R4: (SC 2, inc) x 4 (16)

R5: (SC 3, inc) x 4 (20)

R6: (SC 4, inc) x 4 (24)

R7: (SC 3, inc) x 6 (30)

R8: (SC 4, inc) x 6 (36)

R9-18: SC 36 (36)

R19: (SC 4, dec) x 6 (30)

R20: (SC 3, dec) x 6 (24)

Attach eyes between rows 13 and 14, 5 SC's apart. Use embroidery thread to attach mouth and cheeks.

R21: (SC 4, dec) x 4 (20)

R22: (SC 3, dec) x 4 (16)

Stuff

Change color to dark purple

R23-24: BLO SC 16 (16)

Change color to blue

R25-32: BLO SC 16 (16) -stuff as you go

Change color to dark purple

R33: FLO inc x 16 (32)

R34: BLO SC 32 (32)

R35: dec x 16 (16)

Stuff

R36: dec x 16 (8)

Tie off

LEGS

Part 1 (make 6)

Start with color yellow

R1: MR 4 (4)

R2-6: SC 4 (4)

Tie off

Part 2 (make 6)

Start with color yellow

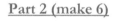

R1: MR 4 (4)

R2-6: SC 4 (4)

Tie off, attack to part 1 at a 90 degree angle. Reinforce with pipe cleaner if desired. Attach to body.

NECK WHISKERS (make 4)

Start with color light purple

Ch 7

Tie off and attach to neck

HEAD ACCENTS

Use light purple yarn to form lines on head

Despite of my Rage I am still just DNA in a phage

DNA

Difficulty: Advanced (difficult sewing required)

Hook: 4.0mm (G) or 3.75 mm (F) hook

Approximate size: 24cm x 8cm

All rows completed in the round unless otherwise indicated

Part 1

Make one with color red, and one with color blue

R1: MR 6 (6)

R2: inc x 6 (12)

R3-62: SC 12 (12) -stuff as you go

R63: dec x 6 (6)

Tie off

Part 2 (make 6)

Start with color red

R1: MR 6 (6)

R2: BLO SC 6 (6)

R3-5: SC 6 (6)

Stuff

Change to color blue

R6-9: SC 6 (6)

Stuff

Tie off

Attachment

Attach as shown below:

For each subsequent attachment, place and attach the ends slightly past, and opposite to where the ends were attached previously (The blue will be to the right of the last blue attachment, and the red will be to the left of the last red attachment). This will create a spiral. Bend and shape project in spiral when finished.

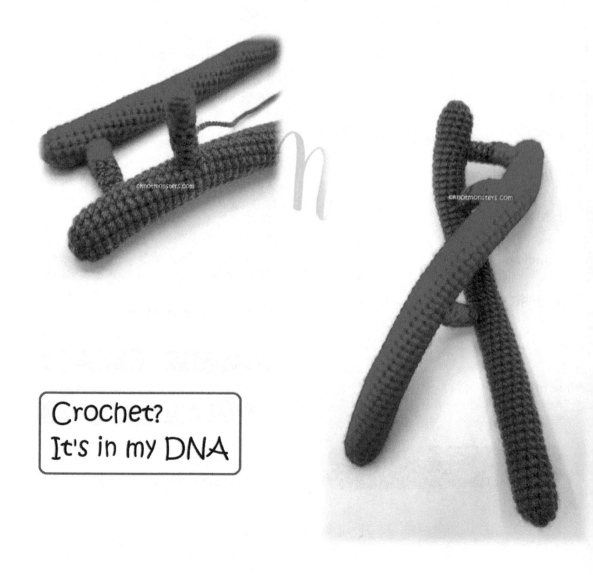

Crochet? It's in my DNA

Virus — COVID 19

Difficulty: Easy

Hook: 4.0mm (G) or 3.75 mm (F) hook

Approximate size: 10cm x 10cm

Eye: 8 mm

All rows completed in the round unless otherwise indicated

BODY

Start with color grey

R1: MR 6 (6)

R2: inc x 6 (12)

R3: (SC, inc) x 6 (18)

R4: (SC 2, inc) x 6 (24)

R5: (SC 3, inc) x 6 (30)

R6: (SC 4, inc) x 6 (36)

R7: (SC 5, inc) x 6 (42)

R8: (SC 6, inc) x 6 (48)

R9-15: SC 48 (48)

R16: (SC 6, dec) x 6 (42)

R17: (SC 5, dec) x 6 (36)

R18: (SC 4, dec) x 6 (30)

Attach eyes, use embroidery thread for mouth and cheeks

R19: (SC 3, dec) x 6 (24)

R20: (SC 2, dec) x 6 (18)

Stuff

R21: (SC, dec) x 6 (12)

R22: dec x 6 (6)

Tie off

Part 2 (make 11)

Start with color dark red

R1: 6 HDC's in a magic ring (6)

R2: BLO (SC, dec) x 2 (4)

R3: SC 4 (4)

Tie off, attach as shown

Attach orange yarn throughout body

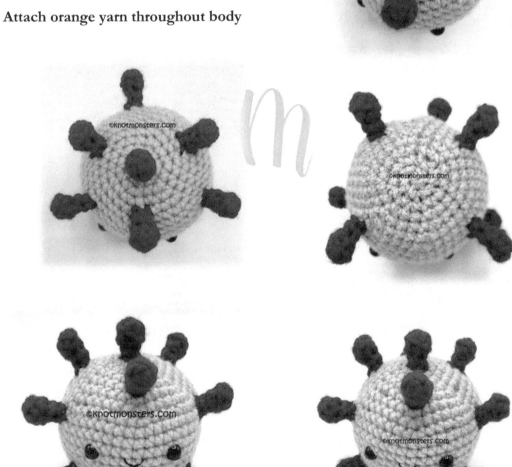

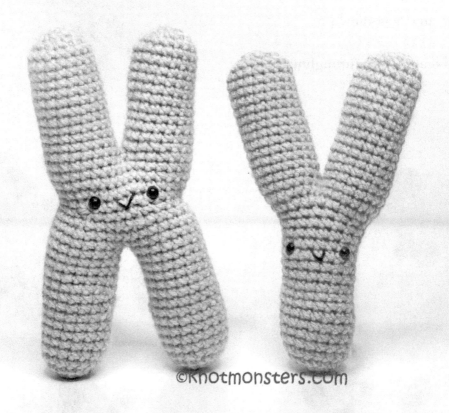

XY Chromosomes

Difficulty: Medium

Hook: 4.0mm (G) or 3.75 mm (F) hook

Approximate size: 15cm x 15cm

Eye: 6 mm

<u>All rows completed in the round unless otherwise indicated</u>

Y CHROMOSOME

Part 1

Start with color blue

R1: MR 6 (6)

R2: inc x 6 (12)

R3: (SC, inc) x 6 (18)

R4-8: SC 18 (18)

Change color to pink

R9-10: SC 18 (18)

Change color to blue

R11-18: SC 18 (18) -stuff as you go

R19: inc x 18 (36)

R20: SC 36 (36)

R21: SC 9, skip 18, SC 9 (18)

Attach eyes, use embroidery thread for mouth and cheeks, stuff. We will be working on one half of the pattern first.

R22-35: SC 18 (18) -stuff as you go

Stuff

R36: (SC, dec) x 6 (12)

R37: dec x 6 (6)

Tie off

We are now going to continue by crocheting the skipped stitches from R21.

R21-35: SC 18 (18) -stuff as you go, tie off loose end from R21

Stuff

R36: (SC, dec) x 6 (12)

R37: dec x 6 (6)

Tie off

X CHROMOSOME

Part 1

Start with color pink

R1: MR 6 (6)

R2: inc x 6 (12)

R3: (SC, inc) x 6 (18)

R4-8: SC 18 (18)

Change color to blue

R9-10: SC 18 (18)

Change color to pink

R11-16 SC 18 (18) -stuff as you go

Tie off, set aside.

Part 2

Start with color pink

R1: MR 6 (6)

R2: inc x 6 (12)

R3: (SC, inc) x 6 (18)

R4-8: SC 18 (18)

Change color to blue

R9-10: SC 18 (18)

Change color to pink

R11-16 SC 18 (18) -stuff as you go

R17: Starting on last stitch from part 1, SC 18, then continue and SC 18 around part 2. You will end with 36 stitches (36)

Tie off loose end from part 1.

R18: SC 36 (36)

R19: (SC 4, dec) x 6 (30)

R20: (SC 3, dec) x 6 (24)

R21: (SC 3, inc) x 6 (30)

R22: (SC 4, inc) x 6 (36)

R23-24: SC 36 (36)

Stuff

Attach eyes, use embroidery thread for mouth and cheeks, stuff

R25: skip 18, SC 18 (18)

We will be forming the 3rd part now.

R26-37: SC 18 (18) -stuff as you go

R38: (SC, dec) x 6 (12)

R39: dec x 6 (6)

Tie off

We are now going to continue by crocheting the skipped stitches from R25.

R25-37: SC 18 (18) -stuff as you go, tie off loose end from R25

R38: (SC, dec) x 6 (12)

R39: dec x 6 (6)

Tie off

If we were chromosomes,
you would be my homologous pair ♡

About the Author

Michael grew up in sunny Las Vegas, Nevada and currently works as a pediatric dentist. His favorite thing in the world is to make others smile, and he hopes his little stuffed creations will bring many more smiles all over the world.

Some fun facts about Michael: His favorite animal is a penguin. Favorite children's movie is "Elf." Favorite sport is roller skating. Favorite crochet pattern is his sunflower pattern. His fondest crochet memory is first starting crocheting and taking three days and many tears to make a crochet ball. A skill that most people do not know about is that he once stacked fifteen Cheerios. If he were a superhero, his superpower would be super speed. He can also recite the general prologue of The Canterbury Tales in old English, which has thus far proved completely useless ☺.

Made in the USA
Las Vegas, NV
12 April 2024

88609243R00044